ERUPTION!

DISCOVERING NEW ZEALAND VOLCANOES

MARIA GILL

NH NEW HOLLAND

CONTENTS

Introduction 3

Chapter One: Our Shaky Isles 6

Chapter Two: Stratovolcanoes 14

Taranaki 15
Tongariro and Ngauruhoe 16
Ruapehu 18
Whakaari/White Island 20

Chapter Three: Shield Volcanoes 22

Mt Cargill 23
Lyttelton and Akaroa 24

Chapter Four: Caldera Volcanoes 26

Rotorua 27
Taupo 28
Tarawera 30

Chapter Five: Volcanic Fields 32

Auckland Volcanic Field 33
Rangitoto 35
South Auckland Volcanic Field 36
Kaikohe-Bay of Islands and Whangarei Volcanic Fields 37
Explosive finds: tuff cones, domes, maars, scoria cones 38

Chapter Six: Submarine Volcanoes and Seamounts 40

Kermadec Islands 41
Waitakere Volcano 41

Chapter Seven: In the Field 42

Monitoring New Zealand's volcanoes 42
Field trips to volcanoes or thermal areas 44
What to do if a volcano erupts 45

Further reading 46

Index 47

INTRODUCTION

Volcanoes are an ever-present threat in New Zealand. They shake and make ash and steam, they can suddenly blow their top without warning, and they can be dangerous to live around. (Does that sound like someone you know?) But they're also hot stuff – and they're exciting!

This book is for all those who want the low-down on our volcanoes: where they are, what type they are, whether they will go off any time soon, and how they may affect you if they do explode. It's a book for people who just love volcanoes or enjoy digesting fiery facts. Perhaps you've visited one of the volcanoes, or are doing a project on one and want to know more about it. Even if you're not studying them yet, they are fascinating once you begin to delve into their craters.

Let me introduce you to Volcanica – she'll be popping up on the pages to help you understand some of the volcano terms in the book.

INTRODUCING VOLCANICA

Hi, I'm Volcanica, the reporter for *Magmatic Press*. I'll be giving you the hard and hot facts about volcanoes in New Zealand. I've visited volcanoes and interviewed scientists around the country, and I've rifled through books, websites and articles to give you the latest information. I've even set up a Volcanic Hotline to answer your volcanic questions on a Facebook page (search: Eruption! Discovering New Zealand Volcanoes).

Welcome to volcanic New Zealand!

SPECIAL FEATURES

Below are some of the extra bits and pieces you'll find in the book:

VIDEOTOPIA This includes film footage and animations for you to look up on the Internet. For example, to build your own volcano online, go to **http://kids.discovery.com/games/build-play/volcano-explorer** and select the 'virtual volcano' option.

MAORITOPIA Get to grips with some Maori legends. For example, the Maori phrase for volcano is ahi tipua. Literally it means 'supernatural fire'. In Maori legend, the gods produced volcanoes, often in order to help humans (or to punish them!).

FACTOPIA Here are some hot facts for you to flash your eyes on. For instance, erupting magma is very hot – often over 1000 °C. This is way hotter than boiling water (100 °C).

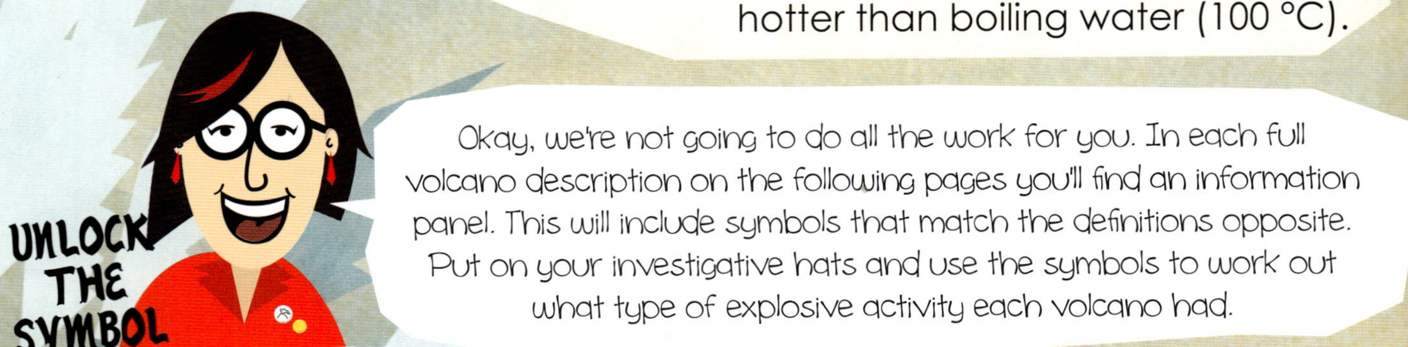

EXAMPLE

Status	Extinct	*Is it active, dormant or extinct?*
Type	Stratovolcano	*What type of volcano?*
Name means	'Seriously loud'	*What does its name mean?*
Altitude	2200 metres	*How high is it?*
Age	25,000 years	*How old is it?*
Last erupted	1897	*When did it last erupt?*
Activities and features	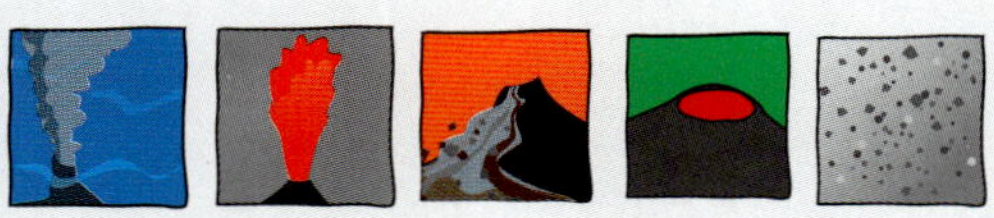	*What type of volcanic activities and features apply to this volcano? See key to symbols.*
Monitoring equipment	web cameras, seismometer microphones, GPS	*What types of equipment are volcanologists using to check on this volcano?*

KEY TO SYMBOLS

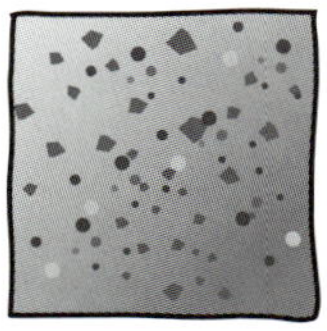	**Ash** Tiny fragments of lava or rock that have been blasted into the air.		**Crater** A cup-shaped vent from which volcanic materials escape.
	**Debris avalanche** A rapid sliding of rocks down a mountain.	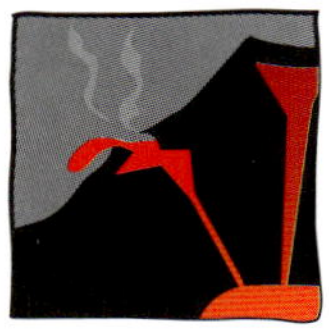	**Fissure** A crack in the side of a volcano through which volcanic materials escape.
	**Fire fountain** A column of frothy hot lava that looks like a fiery red fountain.		**Fumarole** A hole or vent that releases volcanic gases, including steam.
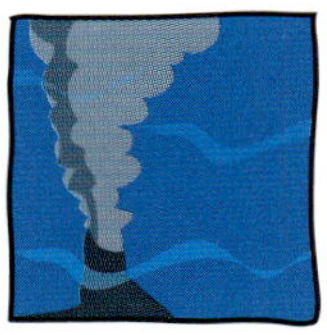	**Hydrothermal vent** A crack in the sea floor from which very hot water escapes.		**Lahar** A river of water, mud and rock debris that pours down a volcano's slopes.
	**Lava flow** Lava that has erupted out of a vent and hardened into rock.		**Pyroclastic flow** A deadly avalanche composed of ash and hot gases that flows quickly down a volcano, hugging the ground.
	**Tephra** Solid material of all sizes (from ash to big blocks) ejected explosively into the air.	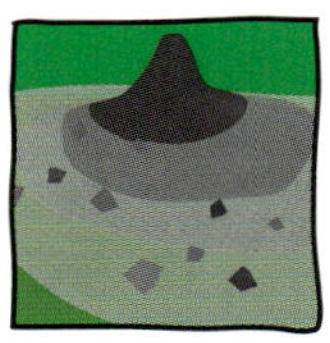	**Volcanic ring plain** The area surrounding a volcanic cone, over which tephra may gather as it settles.

OUR SHAKY ISLES

EARTH-SHATTERING NEWS

We New Zealanders know that the ground beneath our feet is not as solid or stable as it seems. In reality we are standing on huge, slow-moving slabs of rock, called tectonic plates. The plates make up Earth's surface layer. Beneath them is hotter, softer rock, which is constantly being squeezed and reshaped by pressure and heat. This allows the plates to creep about.

The plates are a little like ice floes on water: sometimes bumping together, sometimes drifting apart, sometimes one sliding under another, and other times slipping past another. It is these movements that cause volcanic eruptions and earthquakes along the boundaries where plates meet. Scarily, New Zealand sits astride a plate boundary … in fact, we Kiwis could hardly have chosen a more geologically busy place to live! So before we can explore volcanoes, let's look at tectonic plates.

When plates meet

Not all tectonic plates are the same. The plates that make up the world's major land masses, called continental tectonic plates, are made mostly of granite rock. Those that lie beneath the oceans, known as oceanic tectonic plates, are made up of basalt rock. Basalt is heavier than granite, so the oceanic plates sink lower than the continental plates.

Where a heavy oceanic plate is moving towards a continental plate, it is pushed down beneath its lighter neighbour. This process is called *subduction*, and it is the cause of most of New Zealand's volcanic eruptions.

Why is it that some countries seem to have all the earthquakes and volcanic eruptions? Is it bad luck or is there a reason behind it? Look at these countries or regions on a world map and you'll see that many of them – including New Zealand, Indonesia, Japan and America – form a circle around the Pacific Ocean. Scientists call this circle the Ring of Fire. It is where the boundaries of the oceanic Pacific Plate are being pushed down, or *subducted*, under the lighter continental plates that surround them.

Subduction can happen under the oceans, too. When two oceanic plates are pushed together, one will slip under the other and a deep trench is formed along the boundary. (If you want to find an undersea volcano, look next to a deep ocean trench.)

While plates are subducting along one boundary, they are generally moving apart – or *diverging* – somewhere else. When plates

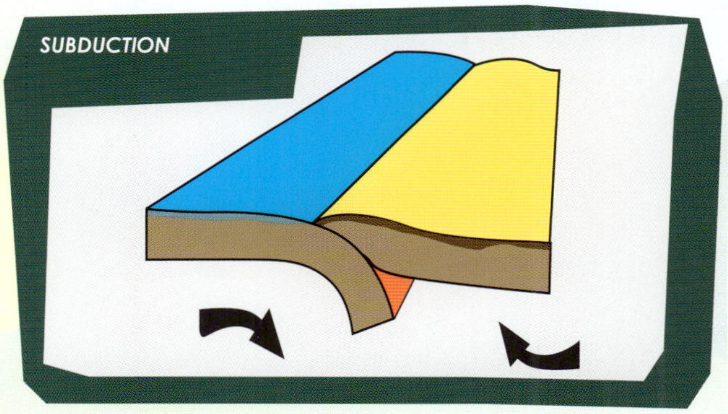

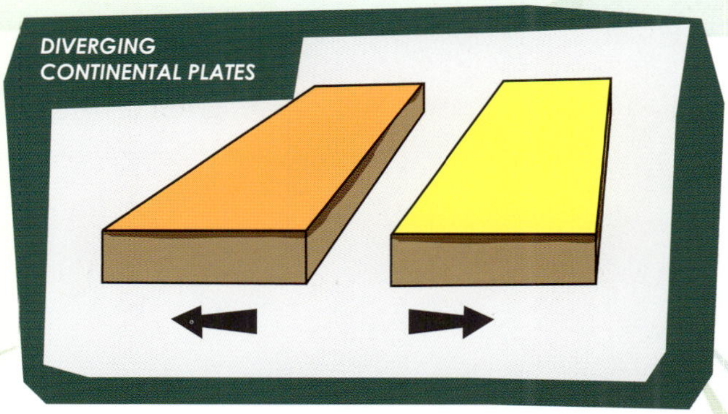

VIDEOTOPIA

Watch the Ring of Fire video: **http://dsc.discovery.com/videos/assignment-discovery-shorts-the-ring-of-fire.html**

Find out more about plate tectonics. Go to **http://dsc.discovery.com/videos/understanding-volcanoes-plate-tectonics.html**

See how plates move. Go to **http://www.iris.edu/hq/programs/education_and_outreach/animations/11**

diverge on land, a rift (crack) opens up. Over millions of years the rift will part the land masses and fill up with sea water. If two plates separate in the ocean, a ridge forms and new seabed is created.

Where two continental plates meet head on, they crumple and push upwards or sideways and create mountains.

Where plates slide past each other horizontally, this is called a *transform boundary* – and this process, too, is at work right now in New Zealand.

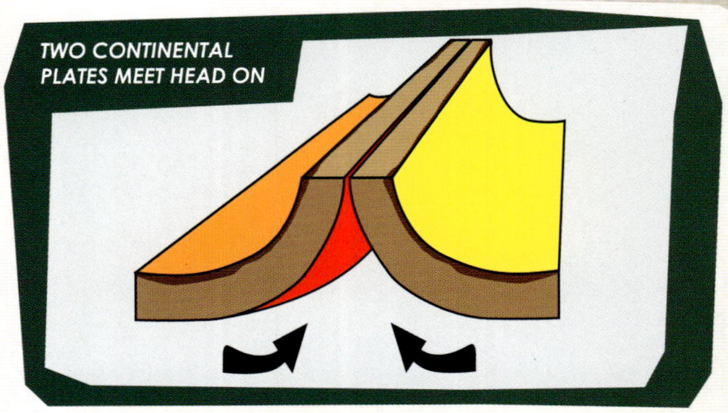

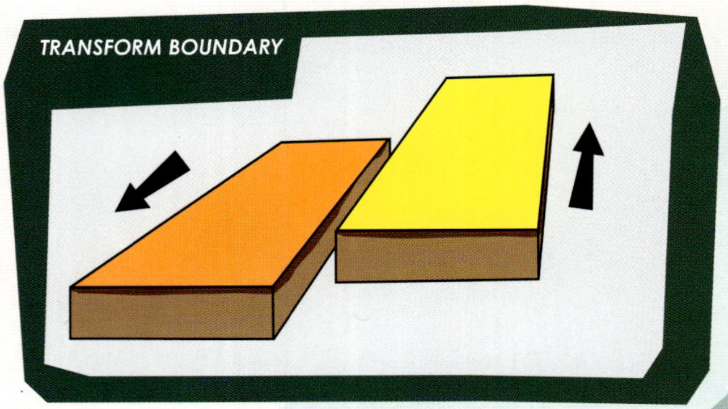

LIVING ON TWO PLATES

New Zealand is volcanically active because it lies astride a boundary between two plates. The North Island and western parts of the South Island sit on the Australian Plate, while the rest of the South Island, Stewart Island and the Chatham Islands are on the Pacific Plate. Just imagine: you might have friends in another town who live on a separate plate!

Along the boundary between the plates, different movements are taking place. To the north-east of New Zealand, the Australian Plate is diving beneath the Pacific Plate. To the south-west, it's the other way round: the Australian Plate is diving eastwards beneath the Pacific Plate. In the middle, the New Zealand land mass is being stretched by all this twisting and diving.

VIDEOTOPIA

Go to **www.youtube.com** and search 'New Zealand under the ocean'.

NEW ZEALAND BEAUTY SPOTS ... MODELLED BY TECTONICS

Some of our country's most spectacular landforms are created by the movement of tectonic plates. Take mountain ranges, for instance: the Southern Alps are being pushed skyward as the Australian Plate squeezes and crumples the edge of the Pacific Plate. And what about that hill? It may be a volcano. Eruptions have thrown up hills as small as One Tree Hill in Auckland or mountains as large as Tongariro and Taranaki. Lakes, too, aren't always what they seem: Lake Taupo, for instance, looks like a calm beauty spot, but it was created by massive eruptions. Smaller lakes could be crater lakes like Lake Pupuke in Auckland. Deep valleys might be hiding a fault line. What occurs on land can also happen under the sea. As you will learn on pages 40–41, New Zealand has some stunning submarine volcanoes and ridges.

MAORITOPIA

Maori used family relationships in their legends to explain the changing landscape of New Zealand. Instead of people, the characters in their stories were gods, fairies and other supernatural beings. This helped them deal with catastrophic events like volcanoes.

The Wellington fault line

ANIMATION

See the different types of volcanoes. Go to **www.iris.edu** and search 'animation – volcanic processes'.

MYSTERIOUS MAGMA

So what exactly *is* an eruption? It is an outpouring of magma, the fiery molten rock from Earth's hot interior. Usually, magma stays below ground, but occasionally it bursts out onto the surface in the form of volcanoes. Volcanoes are explosive proof of the heat from within the planet!

Earth: a really big egg

Scientists believe Earth has several layers, rather like an egg. At the surface is a hard crust – like a shell. (With its jigsaw puzzle of tectonic plates, the crust is like a cracked eggshell.) Beneath the crust is the mantle – a bit like the soft white of an egg. The higher (outer) mantle is cooler and chunkier, like putty and rock. The lower (inner) mantle is very hot. At the centre is the core: this is like the egg yolk, except that the core is made mostly of iron and nickel (and it's very, *very* hot).

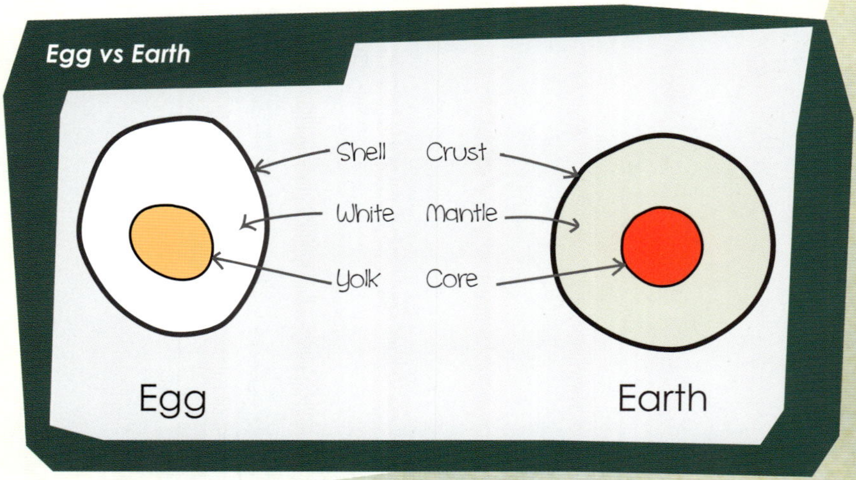

Magma forms when hotter mantle material rises towards the crust, carrying heat with it. As it rises to lower pressures it starts to melt. We call this melted material magma. If you were to imagine the mantle as a sponge, the magma would fill the pores in the sponge. When it is squeezed by the overlying rocks, the magma comes out of the pores and rises to the surface. Think of the blob in a lava lamp rising to the top as it warms. Occasionally, the magma rises all the way to the surface. When it erupts onto the surface it is called lava.

IT'S FIZZICAL

Have you noticed that the deeper you dive in a swimming pool, the more the water pressure makes your ears hurt? Earth is a bit like that: pressure is very high at the core, but much less at the crust. As magma rises, and pressure drops, gases dissolved in the magma are allowed to escape. The gases expand and produce bubbles in the molten rock, which rises to the surface as frothy magma. You can see the same effect if you slowly unscrew a shaken bottle of lemonade. The drink is fizzy only because it contains gas under pressure. By removing the cap, you release that pressure and allow the gas to escape in the form of bubbles.

MAORITOPIA

In legend, Ranginui (Sky Father) and Papatuanuku (Sky Mother) had many children, and these became gods who controlled different areas. When the two parent gods separated, baby Ruaumoko went to live with his mother underground. Ruaumoko controlled volcanoes, earthquakes and geothermal activity. If somebody wanted to take revenge for a wrong done to them they would call upon the gods to help them. Sometimes Mahuika, goddess of fire, would answer, and sometimes Ruaumoko would help.

VIDEOTOPIA

See the convection currents in Earth's mantle. Go to **www.youtube.com** and search 'tectonic plate movement – Auckland Museum'.

Make different types of volcanoes at the flick of a switch!

www.alaskamuseum.org/features/volcano

FACTOPIA

Temperatures in the upper mantle area can be as hot as 800–2000 °C or more – plenty hot enough to turn solid rock into magma.

KNOW YOUR VOLCANOES

Volcanoes come in all shapes and sizes, but they have some common features. Magma rises out of a crack or hole in the ground called a *vent*. The material that erupts out of it slowly builds a volcano. Most volcanoes have a bowl-shaped hole on the summit called a *crater*, which forms when the top of a volcano explodes.

There are four main types of volcano in New Zealand, shown in the table below.

Type	STRATOVOLCANOES	SHIELD VOLCANOES	CALDERA VOLCANOES	VOLCANIC FIELDS
Size and profile	Large steep sides	Medium sloping sides	No slopes	Many small shapes
Shape	Cone	Shield	Crater	Pimple
Crater type	Small crater	Small craters in New Zealand	Large crater	Different sizes
See Page	Pages 14–21	Pages 22–25	Pages 26–31	Pages 32–39
Symbol			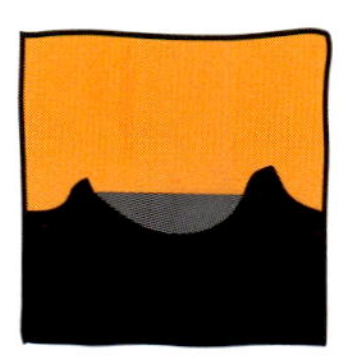	

If a volcano forms in the ocean it is called a *submarine volcano*, see pages 40–41. It may be a *seamount* – completely submerged, like the Brothers Volcano – or have only its tip showing, like Raoul Island. A submarine volcano may also be a stratovolcano or a caldera.

Volcanoes are said to be *extinct*, *dormant* or *active* depending on their state of activity and how long ago they erupted.

A fumarole on White Island

Category	EXTINCT	DORMANT Potentially Active	ACTIVE Erupting
Eruption status	Unlikely to erupt again	Sleeping and may erupt again	Has erupted within last 100 years
Period since last activity	Inactive for millions of years	Inactive for hundreds to thousands of years	Has erupted in the last 100 years or so
Examples in New Zealand	Akaroa and Lyttelton volcanoes	Rotorua caldera, Whangarei and Auckland volcanic fields, Taupo	White Island, Tongariro–Ngauruhoe, Ruapehu, Raoul

STRATOVOLCANOES

Our tallest volcanoes are **stratovolcanoes**, also known as cone volcanoes. 'Strato' comes from a Latin word meaning 'layer'. Stratovolcanoes are formed by many short eruptions in more or less the same place. Some of the eruptions throw out lava and others produce an ash cloud. Over time – maybe thousands of centuries – this material builds up layer upon layer into a cone. Most of these volcanoes have a crater at the top, containing one or more vents.

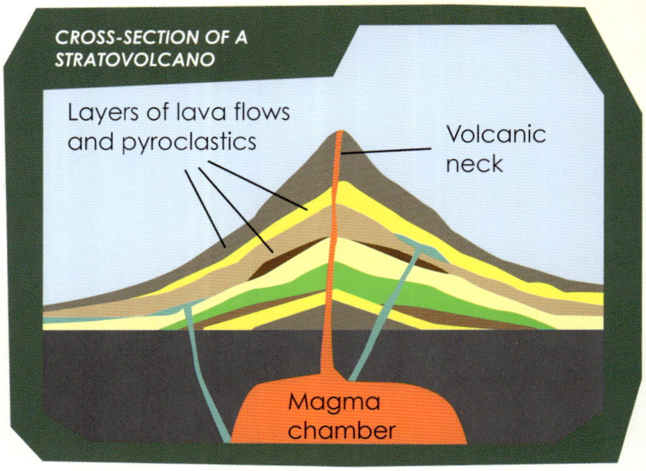

CROSS-SECTION OF A STRATOVOLCANO

Some of New Zealand's most beautiful mountains, such as Ngauruhoe and Taranaki, are stratovolcanoes. Five of our stratovolcanoes are active, but many more are extinct. Some of these are now so weathered, they are hard to spot. For example, did you know that in the West Waikato area there are six extinct stratovolcanoes that erupted nearly three million years ago?

VIDEOTOPIA

See and hear the Maori legend of Taranaki – **http://goo.gl/3oqjc**

FACTOPIA

Local iwi named Taranaki after Rua Taranaki, the first person believed to have reached the summit, in 1420. In 1770, Captain Cook renamed the volcano Egmont, after the Earl of Egmont, First Lord of the Admiralty. In 1978, the volcano's official name was changed back to Taranaki.

MAORITOPIA

In legend, Taranaki was once part of the Tongariro Volcanic Centre, but fought a battle with Tongariro over the beautiful Mt Pihanga. Defeated, Taranaki retreated towards the west coast, carving out the Whanganui River as he went. The mist that surrounds his head and drifts eastward towards Pihanga is a sign of his love for her.

A large *ring plain* surrounds Taranaki

TARANAKI

Status	Dormant
Type	Stratovolcano
Name means	Tara = 'mountain peak', naki or ngaki = 'shining' (referring to its winter snow cover)
Altitude	2518 m
Age	150,000 years
Last erupted	Around 1850
Activities and features	
Monitoring equipment	Web camera, seismometer

LAHAR ALERT

Watch out, climbers! If there's another debris avalanche or *lahar* on Taranaki, you'd better hurry out of the way. In 1998 and 2008, lahars poured down Taranaki's river valleys after heavy rain. In 2010, millions of tonnes of rocks and sand slid down the western flanks and settled at Pyramid Stream, near the Stony River. Past avalanches and lahar deposits created the large *ring plain* surrounding the volcano. Careful climbers avoid Taranaki after heavy rainfall and check with MetService for the weather conditions.

Taranaki is the youngest and largest member of a group of old volcanoes that includes Kaitake, the Pouakai Range, Paritutu and the Sugar Loaf Islands.

Taranaki has had many cone-building periods. Every now and again, the cone has collapsed, triggering a debris avalanche. The part that you see today is only 10,000 years old, having built up mostly after the last major collapse.

The mountain's last big blowout was in 1655 with smaller eruptions in 1775, and it will certainly erupt again. Scientists call Taranaki a dormant volcano. They think the next time it erupts, there will probably be a slow build-up of earthquakes first, giving us a few weeks' or months' warning. In the meantime, people enjoy skiing or climbing on its slopes.

TONGARIRO

Status	Active
Type	Stratovolcano
Name means	'Throwing hot stones'
Altitude	2291 m
Age	250,000 years
Last erupted	2012
Activities and features	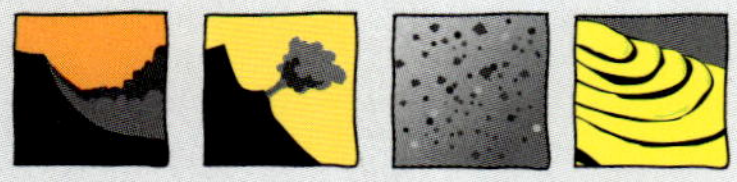
Monitoring equipment	web cameras, seismometer, microphone, GPS stations, checking soil gas and fumaroles regularly

Tongariro might look like a group of volcanoes, but it is actually a single massive stratovolcano. The well-known and very active Ngauruhoe is just one of its vents.

FACTOPIA

In 1887 Chief Horonuku Te Heuheu Tukino IV, paramount chief of Ngati Tuwharetoa, gave Tongariro, Ngauruhoe and Ruapehu mountains to the government. They formed New Zealand's first national park.

VIDEOTOPIA

Watch film of Ngauruhoe erupting: **http://www.teara.govt.nz/en/historic-volcanic-activity/3/3**

Ngauruhoe

Ngauruhoe is New Zealand's youngest and most active volcano. There were over 70 ash eruptions between 1839 and 1975, although it has been sleeping since then. That's an eruption every two or three years on average.

Ngauruhoe's crater has changed many times. In 1841 an ash eruption shortened the top. In 1855 the west side of the crater collapsed, and in 1859 the east wall went, too.

Ngauruhoe erupting in 1975

MAORITOPIA

Many old tales about New Zealand's volcanoes involve a traveller named Ngatoroirangi. He was a tohunga, or priest, of the Arawa people. In one legend, Ngatoroirangi climbed the slopes of Tongariro with his slave, Ngauruhoe. They climbed a big cone volcano, but by the time they reached the snowy summit, Ngauruhoe was dead. The summit was named in her memory.

Magmatic Press

20 February 1975

NGAURUHOE BLOWS!

When Ngauruhoe began discharging lava back in January 1973, little did nearby residents know it was only the beginning of a long build-up to an explosive eruption.

Yesterday, onlookers heard an explosion and saw an expanding ash cloud.

Red-hot lava blocks, some the size of cars, left vapour trails in the sky as they were thrown distances of up to 2 kilometres. The ash plume towered up to 13 kilometres above the crater, before collapsing under its own weight. Ash and scoria avalanches then swept down the sides of Ngauruhoe.

VIDEOTOPIA

Find out about pyroclastic flows and how dangerous they are. Google 'Discovery Channel videos pyroclastic flow'

FACTOPIA

Pyroclastic flows can travel at up to 700 kilometres per hour – that's as fast as a jet!

The black trail of a lahar flowing down from Ruapehu

RUAPEHU

Status	Active
Type	Stratovolcano
Name means	'Exploding pit'
Altitude	2797 m
Age	250,000 years
Last erupted	25 September 2007
Activities and features	
Monitoring equipment	Web cameras, seismometer, microphones, GPS stations

If you visit Whakapapa and Turoa ski fields, you'll be skiing on one of New Zealand's most active volcanoes. Scientists believe Ruapehu began erupting around 250,000 years ago. They have recorded major eruptions every 50 years in 1895, 1945 and 1995–6. There have also been 60 minor eruptions since 1945.

Only a thin layer of ash fell from Ruapehu in 1995–6, but it still created problems for locals. It made their eyes and throats sore, spoiled houses and cars, damaged crops, cut off water and electricity supplies, and prevented aircraft from flying close to the volcano.

VIDEOTOPIA

Watch the 18 March 2007 lahar come tumbling down the mountain. Go to **www.youtube.com** and search 'Mt Ruapehu Lahar'.

Magmatic Press

24 February 1954

DISASTER AT TANGIWAI

Wrecked rail carriages on the banks of the Whangaehu Stream.

Late on 24 December 1953, while most people were preparing for Christmas Day, a trainload of passengers was hurtling towards danger. Bound for Auckland on the Main Trunk Line, they were approaching the Tangiwai rail bridge over the Whangaehu River, near Waiouru. They didn't know Ruapehu's Crater Lake had filled to a dangerous level. At 10 p.m., while the passengers slept, 340,000 cubic metres of water (the same amount of water as in 120 Olympic swimming pools) flooded down the slope. The lahar smashed into the bridge and wrecked it. The train driver couldn't stop the train in time. Four carriages plunged into the river, and another carriage hung over the river until it, too, fell in.

Survivor Joan Karam remembers waking up at 10.30 p.m. and realising that the train was off the rails, lurching and rattling over a rough surface before sinking into the river. 'I felt the splash of cold water through a window,' she recalls, and describes how the water rose till she 'was gasping for breath'. Joan managed to float free and breathe in air 15 centimetres from the ceiling of the train.

'It was of course pitch dark and there was an awful sound around me of rushing water. Both blinded and deafened, I started to feel around me with hands and feet. Two or three feet under the water I grabbed a hand.' She thought it was her sister at first, but realised she had rescued a stranger.

The two young women managed to slide out of a broken window to safety. Joan was discharged from the hospital several days later. She found out that her sister had died in the train wreck, along with 150 other people. The Tangiwai incident remains one of New Zealand's worst natural disasters.

FACTOPIA

Ruapehu has eight glaciers, and they are the only glaciers in the North Island.

WHAKAARI/WHITE ISLAND

Status	Active
Type	Stratovolcano
Name means	'To make visible', referring to the white cloud that always surrounds the island
Altitude	321 m
Age	150,000–200,000 years
Last erupted	2012
Activities and features	
Monitoring equipment	web cameras on island and at Whakatane, seismometer, microphone, regular monitoring of water, gas and soil

White Island sits atop two old stratovolcanoes

Whakaari/White Island is our largest and most active volcano. Its 321-metre high summit, rising out of the Bay of Plenty waters, gives little clue to its true size, as most of the volcano is submerged. Under the water it is 17 kilometres wide!

FACTOPIA

Captain Cook sighted the island on 1 November 1769. He named it White Island because it was shrouded in steam.

VIDEOTOPIA

Watch the volcanologists at work. Go to **www.youtube.com** and search 'scientists on White Island'.

Above the water, most of the island is a large crater with steaming fumaroles and boiling mud pools. White Island has had two episodes of cone growth. It has also had numerous collapses of the crater walls. In 2003, a lake drowned some of the steaming fumaroles.

The island is privately owned by the Buttle Family Trust. They allow tourist companies to take boat, plane or helicopter loads of people to visit the island daily. Scientists monitor the island continuously because it is such an active volcano.

MAORITOPIA

Long ago, when the trickster Maui had fished the North Island out of the sea, he found volcanic fire upon it. Not knowing what it was, he picked it up in his hand. Ouch! In pain, he hurled it into the sea. Where it landed, the volcanic island Whakaari grew out of the waters.

Magmatic Press

11 September 1914

ERUPTION KILLS TEN!

Last night, 10 sulphur miners on White Island were tucked up asleep in their bunk room when the volcano's southern crater wall collapsed, causing a landslide. The miners most likely never knew what hit them as the lahar charged down and swept them and the building out to sea. A rescue vessel was put out, but could find no trace of them. Only the mining company's cat, Peter the Great, survived the disaster.

BOMBS AND BLOCKS

I went undercover to find out about volcanic bombs, thinking lives could be at stake. I discovered that they occur when lava cools and turns solid while still in the air. They are different from volcanic blocks. These occur when lava is so thick it blocks up a vent. Behind it pressure builds up, until the block fractures, and the release throws the block into the air.

A volcanic bomb (top) and a volcanic block (bottom)

SHIELD VOLCANOES

Shield volcanoes – as you probably guessed – look like a warrior's shield. They are not the tallest volcanoes, but they can spew out a huge amount of lava. The lava is not particularly sticky, so it flows freely, like runny honey. It may ooze out of more than one vent, or out of cracks known as fissures. It flows down the slopes along lava channels, which look like rivers of blood, or underground tunnels called lava tubes. (The tubes form as lava flows away from the vent and rapidly cools on the outside to form a solid crust. Meanwhile, the hot lava continues to flow in the inside, eventually draining to leave the hollow tubes.)

Shield volcanoes often spray jet-like liquid lava, like a fountain. These fire fountains can last for short periods or for several hours. Once the lava has spurted into the air it cools very quickly – sometimes it is solid before it even hits the ground. Fallen lava gradually builds up around the vent, making a cone.

Lyttelton and Akaroa Harbours, and Mt Cargill in Dunedin, are extinct shield volcanoes. Rangitoto is also a shield volcano (read more about it on page 35). The Auckland and Whangarei volcanic fields (see pages 33–37) have a few small shield volcanoes, too.

VIDEOTOPIA

See lava flowing from a volcano. Google 'Discovery Channel videos: lava flow'.

FACTOPIA

The best-known shield volcanoes in the world are the islands of Hawaii.

Kilauea in Hawaii is a massive shield volcano, built up from free-flowing 'runny honey' lava

Mt Cargill is a massive shield volcano that rises high behind the city of Dunedin. From its summit you can see volcano-shaped peaks on the Otago Peninsula and north into the Silverpeaks. Scientists believe there are another 200 to 300 extinct volcanoes in the eastern Otago area that erupted from 24 to 10 million years ago.

Before Mt Cargill erupted, much of the land where Dunedin city now stands was under water. The volcano spewed enormous amounts of lava, about 100–150 cubic kilometres. This raised the land above sea level. Scientists have found evidence of 50 lava flow layers in the cliffs at Aramoana.

MT CARGILL

Status	Extinct

Mt Cargill and the remains of weathered extinct volcanoes

FACTOPIA

When molten basalt rock cools quickly it shrinks and cracks into four- to six-sided columns of rock. You can see this columnar basalt at the Organ Pipes on Mt Cargill.

LYTTELTON AND AKAROA

When the explorer Captain James Cook first saw Banks Peninsula in Christchurch he thought it was an island. If he could have time-travelled and sailed there many millions of years earlier he would have been right! It began life as a couple of island volcanoes. Over a long period of time, rivers washed gravel down from the Southern Alps, which built up and connected the islands to the mainland. At the same time, the craters filled with water and became the harbours of Lyttelton and Akaroa. The crater rim forms the Port Hills.

Lyttelton Volcano began erupting about 12 million years ago. Over time, weathering and erosion lowered the cone from 1500 metres to its rather stunted height of 500 metres. The Akaroa shield volcano began erupting 9 million years ago, building a cone 1500–3000 metres high. Its flows buried the lower south-eastern slopes of Lyttelton Volcano (which had stopped erupting a million years earlier). One side of the circular crater of Akaroa broke away, forming the harbour entrance.

Magmatic Press 24 February 2011

EXTINCT IS FOREVER

In Christchurch, people have been worrying that the earthquake activity of 2010–11 will awaken the two volcanoes in Banks Peninsula. Volcanologists say they're extinct and there is absolutely no chance of them erupting again because the plate boundary conditions that caused the eruptions no longer exist. The closest pool of magma is in the central North Island, and the magma cannot travel horizontally that long distance.

Scientists have reassured residents that the ground shaking of the earthquakes had nothing to do with volcanic activity.

The sides of the volcano have breached, creating a natural harbour

LYTTELTON AND AKAROA

Status	Extinct
Type	Shield volcanoes
Name means	Akaroa means 'long harbour'
Altitude	Lyttelton 1500 m, Akaroa 1500–3000 m
Age	12 million years
Last erupted	Became extinct 6 million years ago
Activities and features	
Monitoring equipment	Seismometer to detect earthquake activity

FACTOPIA

Banks Peninsula was a favourite moa-hunting ground for early Maori. By then the volcanoes were long dead.

VIDEOTOPIA

Watch scientists using seismographs and GPS equipment to do a geophysical survey of Christchurch. Go to **www.youtube.com** and search 'Christchurch Geophysical Survey'.

CALDERA VOLCANOES

Calderas are often called supervolcanoes because they are the most powerful and potentially catastrophic of all volcano types.

The word caldera comes from the Spanish for 'cauldron', and it describes the rounded, bowl-like shape of this volcano type.

A caldera volcano occurs when a large chamber of magma gathers below the surface. The pressure builds up in the chamber, then explodes suddenly and violently.

In the biggest eruptions, the ash column can rise as high as 45–50 kilometres. It then crashes back to the ground as superheated clouds of gas and pumice. These rush down and away from the vent as pyroclastic flows, hugging the ground and travelling as fast as 200–700 kilometres per hour. Meanwhile, the ground above the magma chamber collapses, leaving a huge pit, which may later fill with water to form lakes or harbours (if they are on the coast).

CALDERA FORMATION

VIDEOTOPIA

Watch a demonstration of how a caldera collapses. Go to **www.youtube.com** and search 'volcano, how calderas are formed'.

Lake Rotorua only partly fills the caldera; today, the city of Rotorua is built on some of the old lake-bed land.

ROTORUA

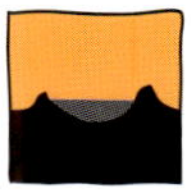

Status	Dormant
Type	Caldera
Name means	'Second lake'
Altitude	285 m
Age	240,000 years
Last erupted	2005 hydrothermal eruption
Activities and features	
Monitoring equipment	Seismometer, GPS stations

THE BIG TEN

New Zealand has 10 caldera volcanoes, all of them either in the Taupo Volcanic Zone or offshore near the Kermadec Trench: Lake Taupo, Raoul Island, Mayor Island, Okataina, Mangakino, Kapenga, Whakamaru, Reporoa, Maroa and Lake Rotorua. The two most active are Taupo and Okataina.

Lake Rotorua may look peaceful, but watch out below! It's a sleeping caldera volcano ...

At the volcano's birth, a massive explosion threw up about 150 cubic kilometres of material, which collapsed to create a huge pyroclastic flow. Later the magma chamber collapsed, leaving a ring around the eastern crater wall and a slope to the crater floor on the western side. For many years afterwards, lava oozed out, making domes on the smoky caldera floor.

Scientists believe Rotorua is dormant and may erupt again.

TAUPO

The largest freshwater lake in all of Australasia is really a supervolcano!

The Taupo Volcanic Zone extends from Ruapehu north beyond Whakaari/White Island. It has several key areas such as the Taupo Volcanic Centre. Underneath Lake Taupo's floor, scientists believe there is a magma chamber only 6–8 kilometres deep.

Since Taupo began erupting about 300,000 years ago, there have been two particularly massive explosions. The first, 27,000 years ago, gave the present lake its shape. So much ash was thrown into the atmosphere, it cooled the climate in this part of the world. Ash was deposited over most of New Zealand, lying tens of metres deep near the eruption vent. After pyroclastic flows spread material far and wide, the magma chamber collapsed, leaving a large hole. Since then, there have been at least 26 smaller eruptions and at the same time a number of volcanic domes formed.

The second really big Taupo event occurred around AD 230. It is considered the largest eruption the world has seen in the last 5000 years. The eruption column may have risen as high as 50 kilometres, and the pyroclastic flow travelled at up to 1000 kilometres per hour as it spread out, covering land within 70–90 kilometres of the vent. The flow was so hot it would have killed everything in its path.

FACTOPIA

On the cliffs of Mine Bay, overlooking Lake Taupo, is a 10-metre high carving that shows the face of legendary priest Ngatoroirangi. Carved in the late 1970s by Matahi Brightwell and John Randall, it is intended to commerate their ancestors and to give the Ngati Rauhoto people pride.

Magmatic Press 20 October 1998

TAUPO – THE LAST FRONTIER

The pilot steered the JAGO mini-sub around the Horomatangi Reefs in the unexplored deepest part of Lake Taupo, reaching a depth of 184 metres. The two-man crew had given up hope of finding anything interesting when they moved up a slope and saw lots of gas bubbles and hydrothermal activity.

The pilot described the 50-centimetre high chimney vents. 'They had white sponges on top – very unusual – like a little lollipop,' he said. 'The area must have been very, very active.'

Until this discovery, Lake Taupo had been the last unexplored geothermal area in New Zealand.

At 35 kilometres across its widest point, the Taupo caldera contains the biggest freshwater lake in New Zealand

TAUPO

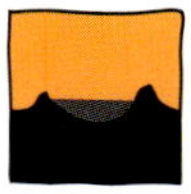

STATUS	Dormant
TYPE	Caldera
Name means	Taupo's full name is Taupo-nui-a-Tia, which means 'the great cloak of Tia'. Tia was an uncle of Tama, the captain of the *Arawa* canoe.
Altitude	760 m
Age	300,000 years
Last activity	AD 230
Types of volcanic activity	
Monitoring equipment	Seismometer, GPS stations, and lake levelling at 22 sites

BLOWN INSIDE OUT

Like me you may be wondering why Taupo isn't a huge mountain; after all, it is our biggest volcano. This is because the eruptions were so big, they hurled all the ash, rocks and lava far from the vent. These volcanic materials filled all the major river valleys in the central North Island with pumice and ash. If you pick up a piece of pumice from a beach such as Whakatane, it most likely came from the Taupo eruption.

VIDEOTOPIA

Find out how dangerous the Taupo Caldera is. Go to **www.qteam.co.nz/episode/4/**

MAORITOPIA

When the Te Arawa tohunga (priest) Ngatoroirangi first arrived in Taupo, he plucked a large totara tree and hurled it into the basin to form a forest. He aimed true, but the wind flipped the tree and it landed upside down, its branches digging holes into the ground. Water rushed up through the holes, filling the basin and creating Lake Taupo.

Close to Taupo Volcanic Centre is another volcanic centre: Okataina. It has produced 6 cones, 12 craters (including two calderas) and 31 domes, including Tarawera.

Tarawera is not the prettiest volcano by any means. It looks like a sandcastle gone wrong! It is made up of three dome-shaped peaks, split down the middle by the largest – and deadliest – eruption to have taken place in New Zealand in the last 500 years.

During the eruption of Tarawera on 10 June 1886, heat from the magma turned the waters of Lake Rotomahana into superheated steam. This cracked the lake-bed rock and caused mud to explode from the dome and settle over a wide area.

Four villages were completely destroyed or buried by falling hot mud, and an estimated 120 people were killed. At Te Wairoa 15 died, but most had survived in stronger buildings. A crater over 100 metres deep had replaced the spot where the beautiful Pink and White Terraces had been, on the shores of Lake Rotomahana.

Magmatic Press

10 June 1886

SLEEPY VOLCANO ERUPTS!

Te Wairoa villagers woke with a jolt at midnight when a series of violent earthquakes shook the ground.

Two hours later, a fissure through Tarawera erupted a cloud of ash and scoria up to 10 kilometres high. At 3.20 a.m. the explosions spread.

'Craters were blasted open on the south-west side of the mountain and through Lake Rotomahana and the Waimangu area. A 17-kilometre rift spewed steam, mud and ash.'

Villagers stood outside to watch Tarawera erupt until they had to retreat to safety. They sought shelter in McCrae's Hotel, until wet mud pelted the roof and it began to fall. Sixty people found safety in the sturdy house of tourist guide Sophia Hinerangi and the Hinemihi meeting house. Three people escaped the local schoolmaster's house when it collapsed, and sheltered in the hen-house.

TARAWERA

Status	Dormant
Type	Caldera
Name means	'Burnt cliffs or peaks'
Altitude	1111 m
Age	400,000 years
Last activity	1886
Activities and features	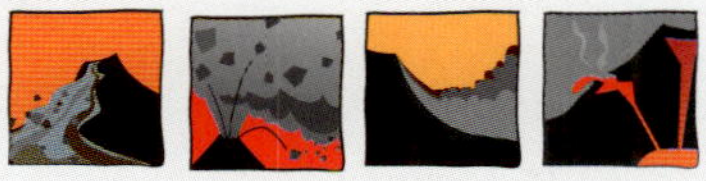
Monitoring equipment	Seismometer, GPS stations, lake monitoring and levelling

The giant rift created by the 1886 eruption can be seen on the right

MAORITOPIA

The tohunga Ngatoroirangi features again in this legend of Tarawera. He went to do battle with Tamahoi, an ogre who lived on the mountain and devoured passing travellers. Stamping on the ground, Ngatoroirangi opened up a huge chasm into which he threw the demon. Then he buried Tamahoi under rock from the mountain. Tamahoi slept for many centuries until he was summoned by another tohunga, Tuhoto. Tamahoi burst out, burying Te Wairoa village in hot molten rock.

VIDEOTOPIA

Hear about Guide Sophia's premonition of an explosion. Go to: **www.youtube.com** and search 'Hei Tiki – Tales from Te Papa'.

VOLCANIC FIELDS

What's a volcanic field? It's what you get when large blobs of magma collect 70–100 kilometres below the surface; occasionally these leak to the surface forming small eruptions over a wide area. The eruptions tends to form small explosion craters, tuff or scoria cones or small lava shields. Each vent usually erupts only once in each place, forming a cone or crater over a period of weeks or several years. It is then considered extinct, although occasionally two eruptions have occurred very close together. A volcanic field typically forms many small scoria cones and craters over hundreds of thousands of years.

New Zealand has several volcanic fields. Auckland is sometimes called the 'City of Volcanoes' because many of its hills and lakes are volcanoes in disguise – there are 55 of them in all! The South Auckland Volcanic Field is even bigger: it boasts 80 volcanic vents. Other volcanic fields include Kaikohe-Bay of Islands, Whangarei, Waikato, and a much older field in Otago.

Magmatic Press

9 November 2011

MORE VOLCANOES IN AUCKLAND!

Scientists have discovered another four volcanoes in Auckland. They include a crater in Boggust Park in Mangere, and three others in Puhinui Reserve in Wiri. This brings the total number of volcanoes in the Auckland Volcanic Field to 55.

The new finds had not been identified before now because they are not major landforms. Fortunately, all four are protected in reserves – unlike many of Auckland's other volcanoes, which have been quarried away or buried beneath housing.

VIDEOTOPIA

See Auckland's volcanoes as they are today. Go to **http://www.youtube.com/watch?v=V6kzgo10_VM**

See how volcanoes are monitored in the Auckland Volcanic Field. Go to **http://www.youtube.com/watch?v=766MRuK08fc**

AUCKLAND VOLCANIC FIELD

Status Active

The Auckland Volcanic Field began erupting about 250,000 years ago. It is considered dormant at present, and because Rangitoto only erupted 550–600 years ago scientists expect the field to erupt again.

There have been 55 volcanic eruptions within a 20-kilometre radius of central Auckland. Nineteen are craters (some with tuff rings), two are shield volcanoes, and the rest are scoria cones (also called cinder cones). Thirty have produced lava tubes (see page 22). Some of the lava caves collapsed when their roofs fell in. And there are fossilised forests, too, where trees were buried under volcanic ash or lava flows. You can see these at Takapuna Reef and near the end of Renton Road near Auckland Airport.

Mangere Mountain in South Auckland

FACTOPIA

Archaeologists have discovered Maori pa, terraces and food storage pits on or near Auckland's volcanoes. Maori also cleared land to grow gardens in the rich volcanic soils. They used the volcanic rock in stone walls or housing foundations. You can still see some of the stone walls at Puhinui near Manukau City.

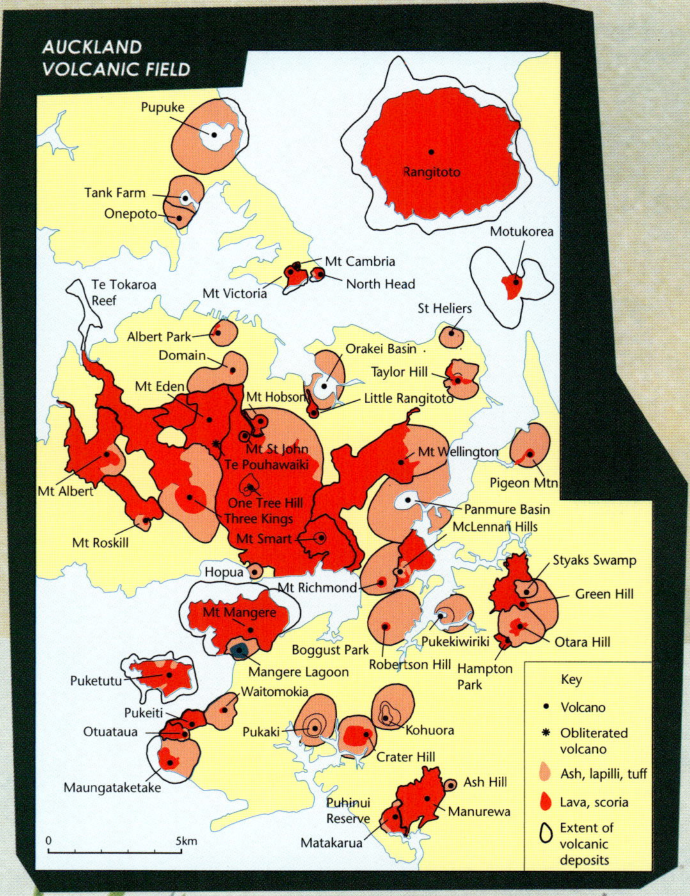

Stewart's Cave is a lava tube left after the eruption of Three Kings Volcano, Auckland

VIDEOTOPIA

Find out about the hot spot under Auckland Volcanic Field. Go to **http://www.aucklandmuseum.com/1207/city-on-the-spot**

Auckland has 50 caves of 20 metres or longer. One has the wonderful name 'Cave of One Thousand Press-ups'. Probably another 50 caves have been destroyed and as many as 50 are yet to be found. Many of the lava tunnels have been discovered by workmen laying pipes. In 2006, workmen found Kitenui Cave in Mt Albert while repairing gas pipes. The workmen said they were digging a hole and it suddenly got deeper and deeper. Later two men crawled through tunnels for about 250 metres and discovered a cavern 20 metres wide. The longest cave in the Auckland area is Wiri cave, which is 290 metres long and up to 6 metres high.

FACTOPIA

Scientists believe some volcanic fields could be active for a million years or more. Auckland's, for instance, has a lot of life left in it!

Rangitoto's eruption was much bigger than any of the other eruptions in the Auckland Volcanic Field

RANGITOTO

Status	Rangitoto: most likely extinct, Auckland Volcanic Field: active
Type	Shield on a volcanic field
Name means	'Bloody sky', which comes from the phrase Rangi-i-totongia-a-Tamatekapua. This describes the battle fought on Rangitoto and the injury given to Chief Tamate-kapua's nose in the fight.
Altitude	260 m
Age	550–600 years
Last erupted	550–600 years ago
Activities and features	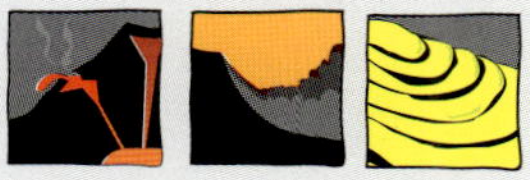
Monitoring equipment	Seismometers, monitoring of soil and gas

Footprints preserved in ash tell us that Maori stood on the shores of Motutapu Island and Auckland to watch Rangitoto forming in the Hauraki Gulf 600 years ago. The fiery event would have beaten any of our modern firework displays! Scientists believe two batches of magma erupted, forming two cones. The second batch spewed lava for several years.

MAORITOPIA

When a giant husband and wife complained that the fire goddess Mahuika let their house fire go out, Mataoho, the god of earthquakes and eruptions, taught them a lesson by sending an eruption to destroy their mountain home. The earth swallowed their home and Lake Pupuke was formed where it once stood. Mataoho then imprisoned the two giants in the peaks of a new mountain surrounded by sea and called it Rangitoto.

FACTOPIA

Rangitoto is home to New Zealand's largest surviving pohutukawa forest.

SOUTH AUCKLAND VOLCANIC FIELD

Status	Extinct

The South Auckland or Franklin Volcanic Field includes Papakura, Drury, Pukekohe and Bombay, and has over 80 volcanic vents. These small vents erupted between about 1.5 and 0.5 million years ago and produced two main types of volcanoes: tuff rings and scoria cones. The field has been inactive for so long that most of the volcanoes have been partly eroded away.

The good thing about having so many volcanoes in one area is that the soil is really fertile. That means it is great for growing those yummy vegetables!

The solid basalt rock around the Hunua Falls is all that is left of this volcano

FACTOPIA

Much of the basalt rock that erupted from volcanoes was used by settlers to make low stone walls separating farm paddocks.

The Hurupaki cone in the Whangarei Volcanic Field

KAIKOHE-BAY OF ISLANDS AND WHANGAREI VOLCANIC FIELDS

Status	Dormant

Kaikohe-Bay of Islands is a small volcanic field with 12 scoria cones near the town of Kaikohe, and three lava flows further east beneath Waitangi in the Bay of Islands. Scientists believe there was another, older volcanic field, 10 million years old, covering a wide area of central Northland between Whangaroa, Kerikeri and Kaikohe.

The Whangarei Volcanic Field may have begun erupting as much as 9 million years ago. Several eruptions created scoria cones near Kamo and north of Onerahi. Unlike Auckland, these eruptions were probably 'dry' – magma most likely did not come into contact with surface water – so they did not start with such a big bang or form a crater.

The cones are nearly eroded away now, but lava flows can still be seen around Whangarei. Another period of eruption around 500,000 years ago produced 10 scoria cones with lava flows. Four scoria cones are west of Kamo and three large scoria cones are west of Whangarei. One small shield cone exists at Whatitiri, west of Whangarei.

Crater Hill is an example of a tuff cone

Tuff cones

A tuff cone has a tall tuff ring, which is a build-up of volcanic debris around a vent in a coastal zone. It is formed when magma touches water and erupts explosively. Examples of tuff cones are Ash Hill, Crater Hill and Shoal Bay in Auckland.

EXPLOSIVE FINDS

In a volcanic field you will find evidence of different types of explosive activity depending on whether an eruption was in a wet or dry area. Discover what scoria cones, maar craters and dome volcanoes have in common.

The dome of Mauao/Mt Maunganui in Tauranga Harbour

Domes

Dome volcanoes are formed when magma rises up out of a vent, like toothpaste squeezed out of a tube, then hardens and plugs the vent shut. Lava domes are found in volcanic fields but also in calderas. Tarawera (see page 31) is the biggest and best-known dome volcano in New Zealand.

A dome volcano you might recognise is Mauao/Mt Maunganui, which towers over one of the most popular beach locations in the Bay of Plenty area. (Locals call it 'The Mount'.) This was the last volcanic eruption in the Coromandel Volcanic Zone – scientists estimate Mauao/Mt Maunganui erupted 4.3 million years ago.

Maars

A maar is a wide round crater found at ground level rather than the summit of a cone. It forms when rising magma mixes with groundwater. This mix is the recipe for a violent explosion, which creates a pit that looks like a meteorite just landed. Because the pit is low-lying, it often later fills up with water. You can see water-filled maars at Lake Pupuke and Orakei Basin in Auckland.

Maar craters (like Lake Pupuke, pictured) and tuff cones are quite similar; the maar crater is only missing the ring around its crater

FACTOPIA

Lava domes often form in calderas – such as Mokoia Island, a dome volcano in Lake Rotorua.

VIDEOTOPIA

Watch a scoria cone erupt. Go to **http://www.youtube.com/watch?v=aS_xl3nu_mY**

Maungatapere is a scoria cone in the Whangarei Volcanic Field

Scoria cones

Scoria cones, or cinder cones, are small volcanoes built over a short period of time, from bits of gassy lava hurled violently into the air. When the magma rose to the surface it most likely touched water, such as swampy ground or a lake, and exploded. With the water blasted away, the magma would have continued to rise, spurting out a fountain of gassy lava, which then hardened to form a scoria cone.

SUBMARINE VOLCANOES AND SEAMOUNTS

A seamount is a mountain that lies completely submerged beneath the ocean. There are some 870 seamounts in the New Zealand region, and most of these are submarine (underwater) volcanoes. In the Kermadec Arc, halfway between New Zealand and Tonga, there are submarine volcanoes bigger than Ruapehu, and chasms deeper than the Grand Canyon. Fifty of these – mostly stratovolcanoes and calderas – have been discovered since the year 2000. Many are active, and many have hydrothermal vents.

SO WHAT'S THE FUSS ABOUT HYDROTHERMAL VENTS?

Hydrothermal vents are a bit like geysers, except that they're on the sea floor, not on land. They pump very hot fluids and smoke into the water. Some are 'black smokers' rich in minerals such as iron, gold and copper. They are home to unusual forms of life such as giant clams and tube worms, which survive on chemicals rather than light. 'White smokers' are cooler than their cousins. Their water contains white minerals such as silicon and calcium. Some smokers are topped with chimney stacks, which build up from metal deposits from the mineral-rich water.

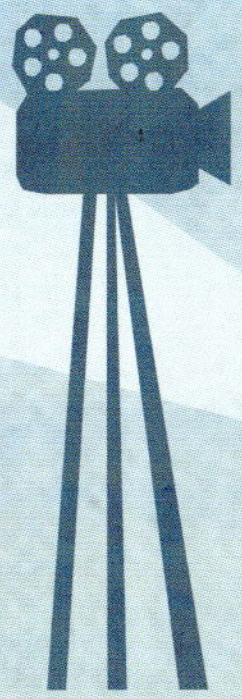

VIDEOTOPIA

See the Kermadec Arc underwater. Go to **www.youtube.com** and search 'New Zealand under the ocean'.

FACTOPIA

Only the Kermadec Islands are above sea level in the Kermadec Arc region.

KERMADEC ISLANDS

The three main volcanoes in the Kermadec region are Raoul, Macauley and Curtis. Most of the volcanoes were originally cone-shaped, but in later eruptions they 'blew their tops', lost a lot of height, and became calderas.

Macauley Island is the visible tip of a giant submarine caldera. The total area of the island is only 300 hectares, but the volcano is nearly twenty times larger.

Brothers Volcano

Brothers Volcano is a seamount that is three times bigger than White Island. It is also a caldera, and on its walls are black smoker chimneys up to 8 metres high. Brothers is part of a chain of seamounts, including the Rumbles (known as I, II, III, IV, and V) that cover a 160-kilometre stretch of the Southern Kermadec Ridge. Rumble III is one of the most active volcanoes in this area.

Brothers Volcano has black smokers with chimneys rising 8 metres high

WAITAKERE VOLCANO

Surfers at Piha Beach would probably be surprised to know that Lion Rock is a volcanic vent that erupted millions of years ago. What's more, the Waitakere Ranges are all that is left of one of New Zealand's largest volcanoes – a submarine stratovolcano that was more than five times the size of Ruapehu and Tongariro! For millions of years this submarine volcano erupted thousands of times from several vents. It built up in size until eventually, around 5–7 million years ago, it rose from the Tasman Sea 20 kilometres west of Piha. Much of it has since been worn away by wave action.

Lion Rock at Piha is an ancient volcanic vent ... and it looks like a sprawling lion

IN THE FIELD

MONITORING NEW ZEALAND'S VOLCANOES

Yikes, run for cover – a volcano is going to erupt! Actually, people usually have more than a few minutes' warning, thanks to volcanologists, who keep a constant eye on all our volcanoes.

Shakes from quakes

One of the warning signs volcanologists look for is earthquake activity underneath a volcano. When magma begins to crack rocks it often causes small earthquakes. These earthquakes can be recognised by the fact that they vibrate at a slower rate than tectonic quakes. They are detected and recorded on an instrument called a seismometer.

If a lahar begins to flow down a volcano's flank, there will be only a few minutes' warning for those below. For this reason volcanologists have a network of instruments installed around Ruapehu's Crater Lake. These will detect signs of an eruption and trigger warnings that alert people on the mountain.

Keeping watch

All of our active and dormant volcanoes have continuous Global Positioning System (GPS) instruments watching over them. The instruments collect radar signals from satellites orbiting Earth to plot their exact location on the planet's surface. This enables them to detect any change or movement on the Earth's surface. GPS devices are very accurate, and some can record changes from one second to the next. Scientists expect to see horizontal and vertical changes of a few millimetres, but anything more than that and they'll be investigating.

Scientists also monitor caldera lakes using a lake-levelling technique, which measures the water depth. They've picked up, for instance, that Tarawera and Taupo have shown changes of several millimetres per year.

They also position webcams (web cameras) that shoot a picture every few minutes of the volcano and its crater. They're broadcast on the Internet, too – which means that if there's an eruption, you can see it happening, almost in real time!

Taking samples

Volcanoes tend to give off gases before and during an eruption. Scientists take samples from fumaroles and crater lakes to detect changes in the chemistry. If it gets quite smelly – like a rotten egg – they know there is more sulphur coming out. They will also collect lava samples to find out what chemicals are in the magma chamber.

Keeping safe

It is quite a risky job walking around active volcanoes so precautions are needed! Volcanologists use safety equipment. They wear a helmet to protect their head from volcanic debris such as tephra and ash; goggles to protect their eyes; gas masks to help them breathe if there is poisonous gas; thick protective clothing in case of falling debris; and big thick boots over their footwear so their feet are not burned.

FACTOPIA

Volcanologists don't just study the volcanoes in their own country. They travel the world getting to know the volcanoes in other countries. They also help the governments of poor countries forecast eruptions and save lives.

VIDEOTOPIA

Look through Geonet's webcams at New Zealand's volcanoes. Go to **http://www.geonet.org.nz/volcano/activity/** and choose a location.

FIELD TRIPS TO VOLCANOES OR THERMAL AREAS

Volcanoes are exciting places to visit, and in New Zealand we have plenty of choice. The smaller cinder cone and shield volcanoes are accessible and easy to climb. To reach Rangitoto or Mayor Island, you need to catch a boat. White Island requires a special trip by boat or helicopter. You can also fly to the stratovolcanoes in the central North Island.

If you go on an organised trip with a tour guide company they will provide safety gear such as gas masks and hard hats, if needed. Whether you go with a tour group or alone, consider taking a waterproof coat, warm clothing, hat and gloves, food and drink, first-aid kit, sunscreen, sunglasses and sunhat. Always tell someone where you are going that day and travel with an adult.

When visiting an active volcano like Ruapehu and Ngauruhoe you need to be aware of the risks. At all times, stay on the path or follow the steps of the guide, and listen to instructions. Enjoy your volcanic adventure!

Climb a volcano on foot

Tongariro Alpine Crossing.
http://www.doc.govt.nz/parks-and-recreation/tracks-and-walks/central-north-island/

Tongariro Northern Circuit (over Tongariro and around Ngauruhoe).
Link as above.

Round the Mountain Track (Ruapehu) at Tongariro National Park.
Link as above, and select 'Ruapehu'

Ruapehu Crater climb.
Link as above. Or, if you want to take the Waterfall Express chairlift, go to **www.mtruapehu.com**

Mt Cargill, Ruapehu, Ngauruhoe, Ngongotaha, Pirongia, Tongariro, Taranaki. (You need to be super-fit for Ngauruhoe and Taranaki, though.)
www.peakbagging.org.nz and search each in turn.

Wairakei Natural Thermal Valley.
http://goo.gl/LjcRD

Travel to a volcano by car

One Tree Hill/Maungakiekie, Mt Eden/Maungawhau, Mangere Mountain, North Head.
http://www.aucklandnz.com/about/nature-and-scenery/auckland-volcanoes

Travel to a volcano by boat

Rangitoto. Travel by Fullers ferry from Auckland City Quay.
www.fullers.co.nz

Mayor Island. Travel by *Enterprise II* from Tauranga.
www.adventurebop.co.nz and scroll down to Mayor Island (Tuhua).

Whale Island. Travel by White Island Tour boat from Whakatane wharf.
www.whiteisland.co.nz

White Island/Whakaari. Travel by White Island Tour boat from Whakatane wharf.
www.whiteisland.co.nz

Taupo caldera. Travel on Great Lake Taupo Cruises from Taupo.
www.greatlakeTaupo.com

Akaroa caldera. Travel on Black Cat Nature Cruise from Akaroa Harbour.
www.blackcat.co.nz

Orakei Korako (geysers and cave).
www.orakeikorako.co.nz

Waimangu Volcanic Valley.
www.waimangu.co.nz

Fly over a volcano

Scenic flights over Rotorua, Crater Lakes, Tarawera volcano, White Island, Orakei Korako Hidden Valley, Tarawera Landing.
www.volcanicair.co.nz

Scenic helicopter flight and landing.
www.vulcanheli.co.nz

See volcanic displays

Auckland Museum Volcanic Forces Centre
www.aucklandmuseum.com

The Volcanic Centre at the Museum of New Zealand Te Papa Tongarewa
www.tepapa.govt.nz

Volcanic Activity Centre in Taupo
www.volcanoes.co.nz

WHAT TO DO IF A VOLCANO ERUPTS

If a volcano erupts, stay indoors, close windows and doors. If you have to go out, wear protective clothing and a mask (or cloth) over your mouth and nose. If you're already outside, find shelter, and cover your mouth and nose.

To discover more about how to plan ahead, and stay safe, visit the following:

'Get Ready, Get Thru'
www.getthru.govt.nz
This safety campaign from the Ministry of Civil Defence gives advice on what to do before, during and after a volcanic eruption, earthquake, tsunami or other major natural disaster.

'What's the Plan, Stan?'
www.whatstheplanstan.govt.nz
Another safety campaign from Civil Defence, this is written for children. You may know it already from your earthquake drills at school. Here you will also find 'Kia Takatu' – safety information written in te reo Maori.

FURTHER READING

Peter Francis and Clive Oppenheimer, *Volcanoes* (2nd edition), Oxford University Press, Oxford, 2004.

Maria Gill and Heather Arnold, *Rangitoto: Te toka tū moana, The rock standing in the ocean*. Penguin Books, Auckland, 2009.

Bruce W. Hayward, Graeme Murdoch and Gordon Maitland, *Volcanoes of Auckland: The essential guide*. Auckland University Press, Auckland, 2011.

Geoff Hicks and Hamish Campbell (editors), *Awesome Forces: The natural hazards that threaten New Zealand*. Te Papa Press, Wellington, 2012.

Lloyd Homer and Phil Moore, *Vanishing Volcanoes: A guide to the landforms and rock formations at Coromandel Peninsula*. Landscape Publications & GNS Science, Wellington, 1992.

Glenys Robertson, *Exploring North Island Volcanoes*. New Holland Publishers (NZ), Auckland, 2005.

Te Ara Encyclopedia of New Zealand, *Life on the Edge: New Zealand's natural hazards and disasters*. Te Ara/Bateman Publishing, Auckland, 2007.

Jocelyn Thornton, *Field Guide to New Zealand Geology*. Penguin Books, Auckland, 2009.

Websites

www.arc.govt.nz/en/volcanoes - volcanoes of Auckland, an educational resource for schools.

www.civildefence.govt.nz – includes information on what to do if there is a disaster.

www.geonet.org.nz – includes information about the volcanoes around New Zealand.

www.getthru.govt.nz – more disaster advice.

www.gns.cri.nz – includes science topics, educational resources and photo gallery.

www.royalsociety.org.nz – includes information about different types of volcanoes.

www.sciencelearn.org.nz – includes volcanic map of New Zealand, types of volcanoes, timeline, videos and simulations.

www.teara.govt.nz/en/volcanoes – explains New Zealand's volcanoes, and monitoring of active volcanoes.

http://volcano.oregonstate.edu – includes volcano legends.

www.whatstheplanstan.govt.nz – includes an educational resource.

INDEX

Akaroa 13, 24–25
ash 5, 28
Auckland Volcanic Field 13, 33–34
Australian Plate 8, 9

block, volcanic 21
bomb, volcanic 21
Brothers Volcano 41

caldera volcanoes 12, 26–31
caves, Auckland's 34
cone volcanoes 14
crater 5, 12

debris avalanche 5
diverging plates 7
dome volcano 38
dormant 13, 16

Earth's layers 10
earthquakes 6, 7, 42
explosion crater 32, 39
extinct 13

field trips 44–45
fire fountain 5, 22
fissure 5, 22
fumarole 5, 13, 21

gases 11, 43
glaciers 18

hydrothermal vent 5, 40

Kaikohe-Bay of Islands Volcanic Field 37
Kermadec Islands 41

lahar 5, 15
lava 10
lava flow 5
lava tube 22, 33, 34
Lyttelton 13, 24–25

maar crater 39
magma 4, 10–11
magma chamber 14, 28
mantle 10
monitoring volcanoes 42–43
Mt Cargill 23
mud pool 21

Ngauruhoe 13, 16–17

Okataina Volcanic Centre 30

Pacific Plate 7, 8, 9
plates, tectonic 6–9, 10
plate boundary 8
pumice 29
pyroclastic flow 5, 17, 26, 28

Rangitoto 35
Raoul 13, 41
ridge 8
Ring of Fire 7
ring plain 5, 15
Rotorua 13, 27
Ruapehu 13, 18–19
Rumbles 41

safety, eruption 45
scoria cone 32, 33, 39
seamount 13, 40–41
seismograph 25
seismometer 42

shield volcanoes 12, 22–25
South Auckland Volcanic Field 36
stratovolcanoes 12, 14–21
subduction 6–7
submarine volcanoes 13, 40–41
supervolcano 26, 28

Tangiwai disaster 19
Taranaki 14–15
Tarawera 30–31
Taupo 13, 28–29
Taupo Volcanic Centre 28
Taupo Volcanic Zone 27, 28
temperature 4, 11
tephra 5
thermal areas 44
Tongariro 13, 16
transform boundary 8
tuff cone 32, 38
tuff ring 33, 38

vent 12
volcanic fields 12, 32–39
volcanologists 20, 42–43

Waitakere Volcano 41
webcams 43
Whakaari/White Island 13, 20–21
Whangarei Volcanic Field 13, 37

To all the volcanologists keeping a watchful eye over the volcanoes for us.

Acknowledgements
Big thanks to volcanologist Dr Gill Jolly from GNS Science for casting her expert eye over the book; she answered all my questions and brought me up-to-date with the latest research on volcanoes. Also thanks to Christine Thomson, Matt Turner, Camille Lowe and the rest of the New Holland team for all the energy they put into the book. I would also like to acknowledge the Buttle Family Trust and White Island Tours for the trip to White Island. It was a great experience to visit a live volcano. **MG**

First published in 2012 by New Holland Publishers (NZ) Ltd
Auckland • Sydney • London
www.newhollandpublishers.com

5/39 Woodside Ave, Northcote, Auckland 0627, New Zealand
Unit 1, 66 Gibbes Street, Chatswood, NSW 2067, Australia
The Chandlery Unit 704 50 Westminster Bridge Road London SE1 7QY United Kingdom

Publishing manager: Christine Thomson
Editor: Matt Turner
Concept design: Yasin Karadeniz
Internal design and illustration graphics: Camille Lowe

Front cover image: Ngauruhoe erupting (image digitally enhanced)
Back cover images: Banks Peninsula (top); Ruapehu (bottom)

National Library of New Zealand Cataloguing-in-Publication Data

Gill, Maria, 1961-
Eruption! : discovering New Zealand volcanoes / Maria Gill.
Includes bibliographical references and index.
ISBN 978-1-86966-362-9 (pbk.)
1. Volcanoes--New Zealand—Juvenile literature.
[I. Volcanoes. 2. Geology.] I Title.
551.210993—dc 23.

10 9 8 7 6 5 4 3

Colour reproduction by Pica Digital Pte Ltd, Singapore
Printed in China by Toppan Leefung Printing Ltd, on paper sourced from sustainable forests.

The photographs featured in this book are by Lloyd Homer, GNS Science, with the exception of the Brothers Volcano on page 41 by Cornel de Ronde, GNS Science, the Tangiwai rail disaster on page 19 from the Alexander Turnbull Library, Wellington, NZ and the photographs of White Island on pages 20–21 supplied by Maria Gill. The map on page 33 is by L. Kermode, GNS Science.